You go into a department store and buy one bath towel for $4.50, two sheets for $19.95 each, a blanket for $24.95, and a shirt for $11.50.

$4.50
19.95
19.95
24.95
+ 11.50
$80.85

Write down the addition problem (with the decimal points lined up) and add to see how much you have spent:

a.

We buy items by exchanging money for them. We say that we are paying for what we buy.
When we pay for an item, we
____ a. exchange money for it.
____ b. promise to buy it later.

The price of an item is the amount of money needed to buy it.

$415.00

$415.00

The price tag on this chair tells us that we would need $_____ to buy the chair.

If you told a sales associate that you wanted the chair, and you gave her $415.00 for it, you would be
____ a. selling the chair.
____ b. buying the chair.

b.

UNIT 1

1

Here are the amounts you have spent on food for the same period:

$139.85
144.99
136.78
140.33
+ 142.65
TOTAL

Find the average for 1 month: _____

You have kept track of your expenses for 5 months. Here are the amounts you have spent on clothing during that period:

$24.01
38.19
44.20
10.45
+ 68.90
TOTAL

Find the average for 1 month. (Divide the total by 5.) _____

In the paper goods aisle, Marie found several different brands of tissues. Marie took box C labeled "Market." It contained more tissues than the other brands.

A
$1.50

B
$1.50

C
$1.50

Was box C the best value? _____

Marie read the boxes carefully. She bought box C because it had the (most, least) amount of tissues for the (highest, same) price.

$39.75	What is the price of this lamp? $_____ Suppose that you buy the lamp. You give the salesperson two 20-dollar bills.
more	You gave the salesperson (more, less) than the price of the lamp.
$0.25	The salesperson must give you: <div align="center">$40.00 −39.75</div>

$0.25	You exchanged part of the two 20-dollar bills for a lamp worth $39.75.
	
	You exchanged the rest of the two 20-dollar bills for a coin worth
	    $0.25
	    $39.75
	
	We call the quarter your change.
	
	If you give a salesperson more than the price of what you are buying, the salesperson must give you the correct change.

yes	You buy a milkshake for $1.45. You give the cashier two dollars. He gives you two quarters and a nickel. Is that the right change? (yes, no)

A budget is a plan for
    ____ a. spending
    ____ b. banking

Sasha Kant lives by herself. She has $400.00 a week to spend. Here is her budget:

FOOD	25%
CLOTHING	20%
RENT	+30%
TOTAL:	%

How much extra does Sasha have to spend? _____%

Sasha's budget allows her to spend the following amount each week on each necessity:

FOOD: $400.00
$$\underline{\times\ .25}$$

CLOTHING: $400.00
$$\underline{\times\ .20}$$

RENT: $400.00
$$\underline{\times\ .30}$$

Food, clothing, and a place to sleep are called
    ____ a. luxuries.
    ____ b. necessities.

$4.22	The total price of your lunch at a cafe is $15.78. You give the person at the cash register a 20-dollar bill. She must give you back: $$\begin{array}{r}\$20.00 \\ -15.78 \\ \hline\end{array}$$
$4.22	The cashier gives you 2 dimes, 2 pennies, and 4 1-dollar bills. How much money did the cashier give you? $____
yes	Did the cashier give you the right change? (yes, no) When the cashier gives you your change, she may count backwards by adding out loud.
$20.00	She will say: "$15.78 and 2 cents (2 pennies) is $15.80; and 20 cents (2 dimes) is $16.00; and 4 dollars is $_____.00."
right	When she adds out loud this way, the cashier is checking to be sure that she gives you the (right, wrong) change.
50 cents **1 dollar** **$5.00**	A woman buys a bag of grapes for $3.50 and gives you, the cashier, a 5-dollar bill. See if you can count the change backwards for her. You will give her 2 quarters and a dollar bill. $3.50 and _____ cents (2 quarters) is $4.00; and _____ dollar(s) is $_____.00.
5 cents **10 cents** **50 cents** **$10.00**	A man buys a pound of fresh tuna for $9.35 and gives you a 10-dollar bill. Give him a nickel, a dime, and 2 quarters, and count the change out for him. $9.35 and _____ cents (1 nickel) is $9.40; and _____ cents (1 dime) is $9.50; and _____ cents (2 quarters) is $_____.00.

UNIT 1

3

$22.77

carefully

$721.00

$879.00

Now Lucas looks at the food average. It is $172.77.

Lucas feels that he has not been taking enough time to shop carefully, and that he does not need to spend so much money on food each month. He decides to budget $150.00 a month for food.

That is $_____ less than the food average.

Lucas hopes his new food budget will make him shop more (carefully, carelessly).

Finally, Lucas looks at the clothing average. It is $53.11. Lucas knows that he does not usually spend that much on clothing in one month. He also knows that once in a while he has to spend much more than $53.11.

For example, in March, Lucas had to buy a new jacket and a pair of sneakers for a total of $150.80.

At last, Lucas decides to budget $25.00 a month for clothes, and to save another $25.00 every month to spend later on more expensive items of clothing.

So now Lucas's budget looks like this:

FOOD:	$150.00
RENT:	521.00
CLOTHES:	25.00
SAVE:	+ 25.00

Lucas has $1,600.00 to spend each month. If he stays within his budget, he will have $_____ left to spend on extra items, or hobbies. Lucas also can save his money for future big expenses, such as college or a new car.

James is making change. Rosa gave him a 20-dollar bill to pay for some groceries.

The price of the groceries is $19.37

James has taken 2 quarters, 1 nickel, and 3 pennies out of the cash register. He counts:

$19.40
$19.45
$19.95

"$19.37 and 3 cents (3 pennies) is $19._____; and 5 cents (1 nickel) is $19._____; and 50 cents (2 quarters) is $19._____."

James's total should have been $20.00

nickel

He has to give Rosa another (nickel, quarter).

We can check the math:

$0.63

```
    1
$20.00   20-dollar bill
−19.37   price of groceries
         change
```

James finally gives her:

$0.63

2 quarters	$ 0.50
2 nickels	0.10
3 pennies	+0.03
Add the change	

yes

Is that the right change? (yes, no)

Lucas' averages were:

FOOD: $172.77   CLOTHES: $53.11   RENT: $521.59

Now Lucas knows about how much he has been spending each month on each necessity.

Lucas thinks about the rent average. It includes the cost of electricity and gas. Lucas knows that the amount he pays for electricity does not change much from month to month, and his rent payment stays the same.

But Lucas' apartment is heated by gas.

more

He would expect to use (more, less) gas in winter than in summer for heating.

Lucas looks at the rent column again:

January	$541.75
February	$550.95
March	$515.92
April	$510.86
May	$505.14
June	$504.92

yes

Did Lucas pay more in the colder months than in the warmer months? (yes, no)

$550.95

So Lucas thinks it would be wise to budget more for rent than the average amount. The highest amount he paid during the 6 months was $_____ in February. To the nearest dollar, that is $551.00 for rent. So Lucas budgets $551.00 for rent.

less

During the warmer months of the year, Lucas expects to spend (more, less) than he budgeted for rent.

When a cashier counts out your change, count to yourself with the cashier. This will help you determine if you are getting the right change.

The price of a portable fan is $18.76. The man who is buying the fan gives the cashier a 20-dollar bill.

The cashier gives the man the following change: 4 pennies, 2 dimes, and a dollar bill.

Count out the change:

4 cents, $18.80

$18.76 and _____ cents (4 pennies) is $18._____;

20 cents, $19.00

and _____ cents (2 dimes) is $_____;

1 dollar, $20.00

and _____ dollar(s) is $_____.

yes

Did the cashier give the man the right change? (yes, no)

If a cashier does not count your change out loud, count it to yourself before you leave the counter.

The total price of Josef's groceries was $39.50. He gave the busy cashier two 20-dollar bills. The cashier dropped 3 quarters into Josef's hand, without counting out loud.

$0.50

Josef checked the change. He should have gotten $._____.
($40.00 – $39.50 = _____)

$0.25

The cashier had given Josef $._____ too much. He gave the cashier back a quarter.

Lucas follows the same procedure for February, March, April, May, and June. Then he goes back through the pages and, on another piece of paper, he writes down the total for each section beside the name of the month.

MONTH	FOOD	CLOTHES	RENT
January	$205.82	$63.98	$511.75
February	160.30	40.05	520.95
March	178.90	150.80	515.92
April	155.16	19.08	510.86
May	162.45	24.76	505.14
June	173.99	19.99	504.92
TOTALS	$_____	$_____	$_____

Fill in the totals for each column.

Lucas now wants to find the average spent each month on food, on clothes, and on rent. Because the totals are for six months, he will divide each total by _____.

FOOD: $6\overline{)1036.62}$

CLOTHES: $6\overline{)318.65}$

RENT:

FOOD: $1,036.62

CLOTHES: $318.66

RENT: $3,069.54

6

$172.77

$53.11

$511.59

$0.23	You give a cashier: $1.00 The price is: − 0.77 Your change is:
$1.32	You give a cashier: $5.00 The price is: − 3.68 Your change is:
$3.56	You give a cashier: $10.00 The price is: − 6.44 Your change is:
$4.25	You give a cashier four 1-dollar bills and a quarter. That is $_____.
$0.09	The price of what you are buying is $4.16. $4.25 −4.16 Your change is:
$0.99	The price of what you are buying is $9.01. If you give the cashier a 10-dollar bill, you will get $._____ in change. ($10.00 − $9.01 = _____)
a.	But if you give the cashier a 10-dollar bill and 1 penny, she can give you ($10.01 − $9.01 = _____) ____ a. a 1-dollar bill ____ b. a 5-dollar bill

In order to make up a budget, you need to keep track of your expenses (how much you spend) over a period of time.

Lucas Santos decides to keep track of his expenses for food, clothing, and rent for six months. He buys a notebook. The first month is January so he writes the following on the first page of his notebook:

Each time Lucas spends money on food, clothing, or rent, he writes the amount in the correct section.

At the end of January, the page looks like this:

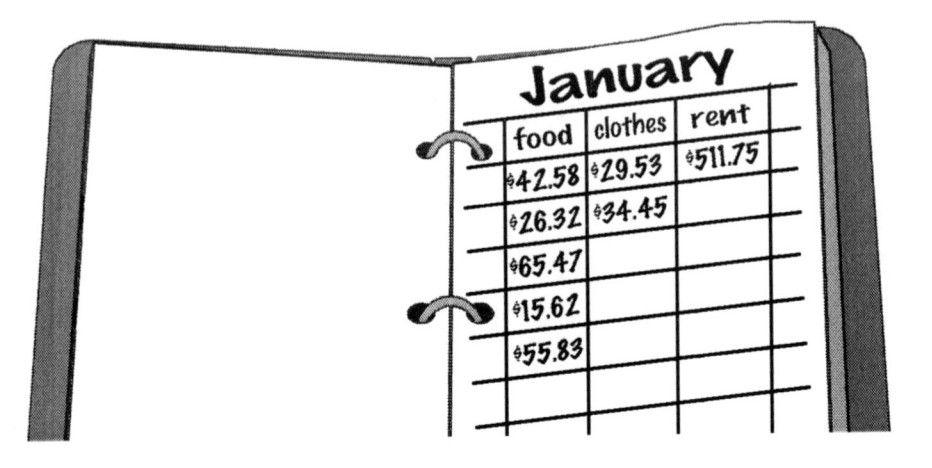

In the rent section, Lucas includes payments both for rent and for utilities (water, electric, telephone, etc.)

Add up each section to find out the total amount Lucas spent on food, on clothes, and on rent.

FOOD = $_____

$205.82

CLOTHES = $_____

$63.98

RENT = $_____

$511.75

$8.00	The price of what you are buying is $7.05. You give the cashier a 5-dollar bill and three 1-dollar bills. That is $___. ($5.00 + $3.00 = _____)
$8.05	The cashier asks if you have a nickel. If you give her a nickel, you are giving her a total of $___. ($8.00 + $.05 = _____)
$1.00	Your change will then be $___ even. ($8.05 − $7.05 = _____)
b.	The cashier can give you ___ a. 3 quarters. ___ b. a dollar bill.
$0.95	If you do not have a nickel, your change will be: $8.00 −7.05

a.	And the cashier will have to give you
	        ___ a. 3 quarters and 2 dimes.
	        ___ b. 2 quarters and 4 dimes.

$3.50	The price of what you are buying is $3.37. You give the cashier three 1-dollar bills and 2 quarters. That is $_____ .
$3.52	The cashier asks if you have 2 pennies. If you give him 2 pennies, you are giving him a total of $_____. ($3.50 + $0.02 = _____)
$0.15	Your change will then be: $3.52 −3.37

Clarence's budget shows that he can spend the following amount each week on each necessity:

FOOD:

$500.00
x .25

$125.00

CLOTHING:

$500.00
x .15

$75.00

RENT:

$500.00
x .30

$150.00

Hadley Carter has $700.00 a week to spend. She is married and has one child. Her budget for food and clothing will be higher than Clarence's.

Find the amount that Hadley budgets for each necessity:

FOOD (30%):

$700.00
x .30

$210.00

CLOTHING (20%):

$140.00

RENT (35%):

$245.00

You give a cashier: **$0.50**
The price is: **– 0.27**
Your change is:

You give a cashier: **$1.25**
The price is: **–1.09**
Your change is:

You give a cashier: **$4.01**
The price is: **–3.51**
Your change is:

You give a cashier: **$10.00**
The price is: **– 9.74**
Your change is:

The price of what you are buying is $4.76. You give the salesperson a 5-dollar bill and a penny. He gives you a quarter in change.

Did he give you the right change? (yes, no)

A woman gives you a 5-dollar bill to pay for a gallon of orange juice that costs $3.28. You give her 2 pennies, 2 dimes, 2 quarters, and a 1-dollar bill.

Count the change:

$3.28 and 2 (pennies) is $_____; and 20 (2 dimes) is $_____; and 50 (2 quarters) is $_____; and 1 (dollar) is $_____.

Was that the right change? (yes, no)

THE END OF UNIT 1

a.

A budget is a plan for
    ____ a. spending.
    ____ b. earning.

We make budgets to be sure we always have enough money to buy the items we need in order to live.

a.

Check the items everyone needs in order to live:
    ____ a. clothes
    ____ b. a TV set
    ____ c. chewing gum

d.

    ____ d. a place to sleep

e.

    ____ e. food
    ____ f. a motorcycle

Food, clothing, and a place to sleep are called necessities.
If you budget carefully, you will always have enough money to buy necessities.

Clarence lives alone. He has $500.00 a week to spend.
Here is his budget:

FOOD	25%
CLOTHING	15%
RENT	30%
TOTAL:	%

70%

more

Clarence tries to spend no (more, less) than the percent of money he has budgeted for each item.

30%

If he budgets well, he has _____%
of his money left for items he wants,
in addition to his necessities.
(100% - 70% = _____)

When you buy an item, you exchange money for it. You try to get the best exchange you can.

Looking for the best exchange, or best buy, is called shopping. You are shopping for a coffee table. A furniture store has one that you like for $174.50.

Instead of buying the table right away, you go to another store to see what its prices are like. The second store has the same table for $172.50.

$2.00

The table costs $_____ less at the second store.

second

You can get the best buy at the (first, second) store.

a.

When we shop, we compare
    ____ a. the prices of the items.
    ____ b. the fronts of the stores.

You decide not to buy the coffee table for $172.50 right away.

You want to do some more shopping. You go to a third store and find another coffee table for $165.00.

This table is different from the others. The wood is not as pretty, feels a little softer, and the legs are shaky.

You decide to go back and buy the table for $172.50.

Now you have compared the quality (value) of two different kinds of coffee tables.

more

You chose to pay a bit (more, less) for better quality.

UNIT 2

9

Grocery stores usually have ads in the Sunday newspaper. Prices of food may change from week to week, so they list **specials** to let buyers know of the different deals on food.

A **special** is a sale.

You will pay (more, less) than regular price for an item you buy on special.

less

Preeya sees in a grocery store's ad that portabello mushrooms are on special for $.94 a pound. This is not the store where Preeya usually shops, because it is 10 miles from her house, and gas prices are high.

Preeya decides to take the bus to the store so she also can save money on gas. Her bus fare is 95¢ each way, or a total of $_____.   (2 x 95 = _____)

$1.90

Preeya buys 4 pounds of the portabello mushrooms. How much does she pay? $_____

$3.76

Portabello mushrooms in the store near Preeya's house, where she usually shops, are priced at $1.20 a pound. If Preeya had bought them there, she would have paid $_____.

$4.80

Preeya thinks she saved $1.04 by going to the store where the mushrooms were priced at $0.94 a pound.

However, she has forgotten about the bus trip just to buy the mushrooms. The bus fare would be added to the price of the mushrooms: $3.76 + $1.90 = $_____

$5.66

Preeya really lost _____¢. ($5.66 - $4.80 = _____)

86¢

To be thrifty is to spend money wisely.

Was it thrifty of Preeya to go 10 miles by bus to buy portabello mushrooms on special? (yes, no)

no

When you start to shop for an item, or product, you must first decide how much money you can afford to pay.

Dario has saved up $300.00 for a new suit.

He has tried on two suits, and he likes them both. Both suits fit him well and they are stylish. Dario checks the fabric of each and finds both to be a good quality. He looks at the buttonholes, buttons, and pants' zipper. These too, are all well crafted.

He decides that both suits are of the same quality.

One suit costs $298.50. The other costs $292.95.

price

There is a difference in (quality, price) between the two suits.

$292.95

It seems to Dario that the ($292.95, $298.50) suit is the better buy.

no

Did Dario spend more than he could afford? (yes, no)

When we shop, we compare price and quality.

best

We do this to be sure that we get the (worst, best) exchange for our money.

budgeting	When you make a plan for spending, you are (buying, budgeting).
shopping	When you compare quality, quantity, and cost before buying, you are (selling, shopping).

If you live in a place where there are many grocery stores from which to choose, you will help keep the food budget down by comparing different stores' prices.

An easy way to do this is to check the grocery stores' prices and weekly sales in the Sunday newspaper.

Crestline Market	**Clover Market**
$1.20 lb	**$1.48 lb**

We know that Clover Market has lower prices on frozen foods. So, if you have the time, you may want to go to the Crestline Market for fresh carrots and to the Clover Market for frozen foods.

$2.40

For example, you will pay $_____ for 2 pounds of carrots at the Crestline Market.

$2.96

At the Clover Market, the same quantity of carrots would cost $_____.

87

UNIT 12

might	Jean needed a new dress. She decided to spend no more that $100.00. She found two cotton dresses she liked. They both seemed to be well made, but one cost $89.50 while the other cost $97.00.

Jean read the label information. The label on the $97.00 dress said "guaranteed not to shrink." This told her that she could wash it and it would still fit.

The label on the $89.50 dress did not say this. This told Jean that the $89.50 dress (might, might not) shrink after washing.

Jean decided that a dress that might shrink was not a very good buy. |
| **$97.00** | She bought the dress that cost ($97.00, $89.50). |
| **$10.00** | Nina tried on two dresses. One cost $73.00 and the other cost $63.00.

Nina could have afforded the $73.00 dress, but she decided to save $_____ by getting the $63.00 dress. |
| **yes** | Did Nina compare prices? (yes, no) |
| **no** | Did she compare quality? (yes, no) |

$$\frac{1}{4} = \frac{25}{100}$$

.25

25%

less

a.

A **budget** is a plan for spending.

Suppose you are married and have one child. You have $2400.00 a month to spend, and the monthly food bill comes to about $600.00.

$600 is 600/2400 of the total amount you have to spend in a month.

$$\frac{600}{2400} \implies \frac{}{4} \times \left(\frac{25}{25}\right) = \frac{}{100}$$

25/100 ⟹ ._____

.25 ⟹ _____%

We can say that you spend 25% of your money on food.
If you plan your spending so that you do not spend more than 25% of your money on food, we say you have **budgeted** 25% of your money for food.

This means that you will try to spend no more than 25% of your total income.

But you will probably also try to spend even (less, more).
To keep your food budget down (spend 25% or less), you must shop carefully.

You may live in a place where there is only one major grocery store. You will shop for food in that store by comparing
____ a. quantity, quality, and cost.
____ b. the colors of different labels.

smaller	Some clothing, when washed in hot water or machine dried on a hot setting, will shrink and/or fade.
	To shrink is to get (smaller, larger).
color	To fade is to lose (style, color).
cold	Most clothing has labels that tell us how to launder (wash) it. If a shirt may shrink or fade, the label will say "wash in (hot, cold) water."
shrink **lose**	If you wash it in hot water, it may (shrink, stretch) and (keep, lose) some color.
more	An article of clothing that will not shrink or fade is a better buy than one that will, even if it costs a little (more, less).

The labels on clothing in the stores are there to help you compare quality. They tell you what the cloth is made of and how to clean it.

Here is the label from a blouse.

HAPPY FABRICS
100% POLYESTER
WASH BY HAND IN
COLD WATER.
DO NOT DRY-CLEAN

polyester

The blouse is made of (cotton, polyester).
100% polyester means that it is all polyester.

wash

The label says to (wash by hand, dry-clean) the blouse.

Priced at		Cost for 1
3 cans for 98¢	⟶	_____ ¢
7 cans for $5.00	⟶	_____ ¢
4 cans for $3.83	⟶	_____ ¢
2 cucumbers for 75¢	⟶	_____ ¢
3 limes for $1.65	⟶	_____ ¢

Summer squash is selling for 72¢ a pound. How much will you pay for 2.4 pounds? _____

There are _____ eggs in a dozen.
Which cost more?
_____ a. medium eggs
_____ b. small eggs

Which costs more:
_____ a. the white bread
_____ b. the french bread

WHITE BREAD
$1.29

FRENCH BREAD
$1.59

You have tried on two pairs of khaki pants. You like them both and they cost exactly the same amount.

Here are the labels you find on them:

pair A   pair B

cotton	Pair A is made of (cotton, wool).
no	Does the label for pair A say that the cotton has been treated so that it will not shrink or wrinkle? (yes, no)
preshrunk	Read the label for pair B. It is also made of cotton. However, the cloth was shrunk before it was cut and sewn, so it will not shrink again. The word that tells us this is (khakis, preshrunk).
yes	Were khakis B treated so they have less chance of wrinkling? (yes, no)
do not need	The label for pair B also says that the khaki pants (need, do not need) to be ironed.
pair B	Since both pairs of khaki pants cost the same, your best buy is probably (pair A, pair B).

There are many choices to make at a grocery store. It is helpful to prepare a list of what you need before you go. This way, you will not forget an item you might need during the week. Also, a shopping list helps you avoid buying foods or products you do not need. Try to avoid going shopping when you are hungry. You may buy a lot more food than you could possibly eat!

Here is a list Ravi took to the store.

gallon of milk
dozen eggs
4 oranges
white bread
6 tomatoes
lettuce
2 cans of soup

4

How many oranges does Ravi want?

At the store, Ravi saw loose oranges, sold by the count, for 66¢ each, and a 4 lb bag of oranges for $2.20. Ravi could see there were 4 oranges in the bag, and they were the same size as the loose oranges.

66¢

Ravi figured that if he bought 4 loose oranges at _____¢ each,

$2.64

he would pay $_____ for what he needed. Ravi bought the

44¢

pre-packed bag and saved _____ .

When you shop for clothes, you try them on to see if they fit. You look at the way they are made, and then you read the labels.

quality

These are factors to help you consider the (quality, price).

yes

Since you are trying to find the best quality for the money you can afford to spend, should you also compare price? (yes, no)

When you get very high quality for a low price, you have a bargain.

Hank Rosman went into a men's store to buy a shirt. The sales associate showed him a shirt for only $9.98 and said, "This is a real bargain."

Hank bought the shirt. The first time he washed it, it shrank, some of the stitches broke, and the color faded.

no

Was the shirt a real bargain? (yes, no)

quality

Hank forgot to check the (quality, price).

35¢

Cucumbers are marked 2 for 69¢.

1 cucumber would cost _____¢.

33¢

Ears of corn are marked 6 for $1.98.

1 ear of corn would cost _____¢.

---

b.

Eggs also are sold by count. They are sold by the dozen (12).

Eggs are sorted according to size. This is because
   ____ a. all eggs are the same size.
   ____ b. eggs can be different sizes.

You can buy a dozen small, medium, large, or extra large eggs. Usually the price of large eggs is a little more than the price of small eggs.

Most people buy eggs according to the size they like best and find most useful in cooking.

Eggs also can be sold by the ½ dozen or the 1½ dozen. 1½ dozen equals _____ eggs.

18

---

more

For people watching their weight or cholesterol, containers of egg whites or egg substitutes can be found in supermarkets. This is a processed food, (manufactured, not natural) so it will cost (less/more) per ounce than fresh eggs.

It is not easy to find bargains. The best advice to you is to shop carefully and try to get the most for your money.

This means that you will try to get the (best, worst) quality for the price you can afford to pay.

best

No matter what you are buying, it is always smart to shop carefully.

Frida Humes bought a hammer for $5.49. Within six months she had snapped off part of the claw on the back of the hammer.

Was the hammer of very high quality? (yes, no)

no

Finally, Frida threw the hammer away and bought another one just like it.

She had spent $_____ on hammers in six months.
(2 x $5.49 = _____)

$10.98

Taylor Barns bought a more expensive hammer for $9.99.

Taylor used the hammer often and yet, at the end of a year, the hammer was still in excellent shape.

Who shopped more carefully?
____ a. Frida Humes
____ b. Taylor Barns

b.

Some kinds of produce are not sold by weight, but by count.

no

If limes sell for 3 for $1.58, are they sold by weight? (yes, no)

53¢

At this price, one lime would cost _____ ¢
(You have to pay an extra cent.)

Grapefruits are selling at 2 for $2.99.

$1.50

One grapefruit will cost $_____.

Small fruits, like berries, are often priced to sell by the basket.

$4.99

The sign over the strawberries means 2 baskets for $_____.

$2.50

One basket would cost $_____.

most

When you buy fruit in baskets, it is a good idea to examine the baskets. If the strawberries in one of the baskets do not look fresh and ripe, you would be wise to buy a basket of strawberries that look the (most, least) appetizing.

The quality of an object is
   ____ a. how much it costs.
   ____ b. how good it is.

A real bargain is
   ____ a. low quality for a high price.
   ____ b. high quality for a low price.

Your best buy is
   ____ a. the best quality for the amount that you can afford to spend.
   ____ b. the best quality no matter how high the price.

Labels on clothes help us to compare (quality, color).

Here is a dress label:

> # 100% Cotton
> ## No Ironing Needed
> ## Machine Wash
> ## Guaranteed NOT to Shrink

Check the details the label tells you about the suit:
   ____ a. The dress is made of wool.
   ____ b. The dress is made of all cotton.
   ____ c. The cotton is treated not to shrink.
   ____ d. The dress must be dry-cleaned.
   ____ e. The dress can be washed in a washing machine.
   ____ f. The dress is made of nylon.
   ____ g. The dress does not need to be ironed.

You can learn to tell good quality by
   ____ a. shopping and comparing.
   ____ b. buying the first item you see.

THE END OF UNIT 2

7 lb	The bag on the scale has potatoes in it.
	The potatoes weigh _____ lb.
	The potatoes are marked to sell at 2 pounds for $1.00.
50¢	That is _____¢ a pound.
$3.50	So this bag of potatoes will cost $_____.

	Melon is selling for 99¢ a pound.
4 lb	The melon on the scale weighs _____ lb.
$3.96	So it will cost $_____.

	Onions are priced at 75¢ a pound. You weigh out 4 lb 8 oz of them.
4 1/2, 4.50 lb	That is 4 _____ /2, or 4._____ lb.
$3.38	How much will you have to pay? $_____

Dave's friend Dimitri paid $550.00 for a stereo system. He bought it at his favorite store, without comparison shopping.

Dave wanted to get a stereo system just like Dimitri's, but he decided to see if he could get a better buy.

Dave shopped in three stores. He finally found a store that sold the same stereo system for $475.00.

$75.00

Dave saved $_____ by taking the time to visit different stores.

Here are two saucepans. They are **identical** (exactly the same).

Store A   Store B

yes

Does this mean that they are of the same quality? (yes, no)

We find out that the saucepans are for sale in two different stores.

Store A sells the saucepan for $29.99.

Store B sells it for $31.69.

no

Does a higher price always mean higher quality? (yes, no)

---

comparison shopping

Looking for the best buy is called (comparison shopping, paying).

You have $35.00 to spend on a shirt. You find two shirts of the same quality, and both are priced under $35.00. One shirt, however, costs a bit more than the other.

Your best buy is
    ____ a. the higher-priced shirt.
    ____ b. the lower-priced shirt.

b.

The apples cost $_____ a pound.

ROME APPLES
$1.39/LB

Your bag of apples weighs 5½ pounds.

Find the price of the apples: $_____

Oz stands for (ounces, pounds).

Potatoes that weigh 2 lb 8 oz weigh 2½ lb, or 2._____ lb.

When you multiply to find a price and there are more than 2 decimal places in your answer, will you have to pay an extra cent? (yes, no)

POTATOES
97¢/LB

If the potatoes cost 97¢ a pound, find the price of 2.5 pounds. (Give your answer in dollars and cents.) _____

A plain coat has a tag "100% cotton; dry clean only; $94.50."

A coat with pockets has a tag "synthetic fiber, machine washable, no ironing; $68.50." (Synthetic means man-made fabric.)

Desiree Wilson is shopping for a lightweight coat for under $100.00. She is considering these two coats. She tries on each one. They both fit, and both seem to be well made. She looks at the tags.

no	The plain coat is made of cotton. Does its tag say it can be washed? (yes, no)
is not	The pocketed coat (is, is not) made of cotton.
yes	Can it be washed in a machine? (yes, no)
no	Does it need to be ironed? (yes, no)
	Look at the prices.
more	In this case, the coat that costs (more, less) seems to be the better buy.

yes	When you divide to find a price, and you have a remainder in your answer, will you have to pay an extra cent? (yes, no)
	Find the cost of 1 can:
51¢	3/$1.52
64¢	8/$5.09
68¢	6/$4.05
99¢	4/$3.95
$1.23	9/$10.99
66¢	5/$3.29

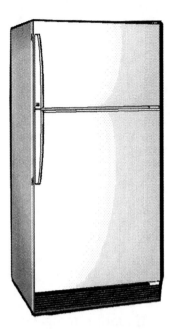

BIG BARGAINS!!!
LONG TERM
WARRANTIES!

Appliances (stoves, refrigerators, washing machines, etc.) and electronic equipment, like DVD players and TV sets, often have **warranties.**

A warranty protects the buyer. In a warranty, the company that made the appliance promises to fix it if it breaks within a certain period of time. A warranty may last for just a few months, or for as long as five years.

Hector bought a second-hand refrigerator for $380.00. There was no warranty with it. At the end of six months, the refrigerator broke down. Hector had to pay $230.00 to get it fixed.

$610.00

In six months, Hector had spent a total of $___ on the refrigerator.

Hector's brother, Miguel, bought a new refrigerator with a two-year warranty. He paid $600.00 for it. At the end of a year, the refrigerator stopped working. The company that made it had to fix it. Miguel did not have to pay for the repairs.

After a year, he still had spent a total of only $600.00 on the refrigerator.

less

That is (more, less) than Hector spent in six months.

Miguel

Who do you think shopped more carefully? (Hector, Miguel)

2 4/16 lb

2 1/4 lb

2.25 lb

The deli attendant weighs some provolone cheese for you. It weighs 2 lb 4 oz.

That is 2 ___/16 lb

or 2 ___/4 lb

or 2.___ lb

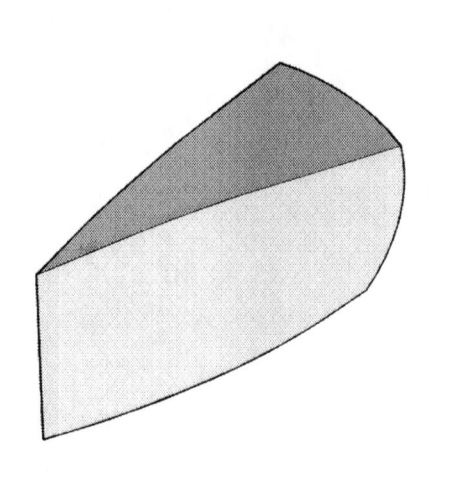

$11.01

The cheese costs $4.89 a pound.

Find the price of the provolone cheese: $_____

(Do not forget that you will have to pay a whole extra cent if your answer does not come out in even cents.)

---

yes

When you have multiplied cents or dollars and cents, and there are more than 2 decimal places in your answer, will you have to pay an extra cent? (yes, no)

$3.09

Containers of cole slaw are labeled 2/$3.09. This means that you can buy 2 containers for $_____.

Find the price of one container:

1.54 R1

$$\overset{\text{R \_\_\_}}{2\overline{)\ \$3.09}}$$

$1.55

The remainder tells you that you will have to pay an extra cent. So 1 container of cole slaw will cost $_____.

.25

.06

15% ⟵ This is a percent.
We can write it as a decimal: 15% ⟹ .15
30% ⟹ .30
25% ⟹ ._____
6% ⟹ .0_____

European Imports, a furniture store, is having a sale on imported furniture. Here is part of the store's ad in the paper:

> Armchairs....................regularly $210.00
> End tables....................regularly $ 70.00
> Couches.....................regularly $220.00
>
> 15% DISCOUNT ON ALL ITEMS LISTED!

To find the amount of discount on the couch, we change the percent to a decimal and multiply:

$$\begin{array}{r} \$220.00 \\ 15\% \Longrightarrow \quad .15 \Longrightarrow \quad \underline{\times \ .15} \end{array}$$

$33.00~~

There are 4 numbers to the right of the decimal point in the problem. Move the decimal point in the answer 4 places to the left. This is the amount of discount. To find the sale price, we subtract the amount of discount from the regular price:

$$\begin{array}{r} \$220.00 \\ \underline{-33.00} \end{array}$$

$187.00

A **discount** is an amount of money subtracted from the regular price.

Items sold at a discount sell for (more, less) than the regular price.

less

UNIT 3

Suppose you want to buy some seedless grapes. The grapes are in clear plastic bags and are different weights. You choose a bag, put it on the scale, and find that the package weighs 1 lb 12 oz.

1 3/4 = 1.75

1 lb 12 oz = 1 _____/4 lb = 1. _____lb

The price per pound is $0.99. We can calculate the total price:

$1.75
x 0.99
$1.7325

$$\begin{array}{r} \$1.75 \\ \times\ 0.99 \\ \hline \end{array}$$

The answer, $1.7325, is not in whole cents. The 25 is a remainder in cents. It means that the answer is more than $0.73 by part of a cent and should be rounded to the higher cent, 74.

You will have to pay a whole extra cent.

$1.74

So your 1 lb 12 oz of grapes will cost $_____.

You decide to buy some sliced swiss cheese from the deli. The deli attendant weighs the cheese and comes up with a price of $5.7725.

$5.78

You will have to pay ($5.77, $5.78).

You can sometimes find good buys by shopping at discount stores, where almost all items are sold at less than the regular price. You can compare the prices in a discount store with the prices in other stores to see how much you save.

If you shop in discount stores, you should check quality very carefully.

Lin Chea bought a chair in a discount store. She paid $198.50 for it. Lin thought that the chair was exactly the same as a chair she had seen in a regular store for $210.00

$11.50

Lin thought she had saved $_____.

However, Lin had not checked the quality very carefully. In a few months, most of the springs in the chair were broken. The cloth was worn in spots, and one of the arms was so loose that Lin was afraid it would fall off.

no

Did Lin really get a good deal? (yes, no)

Gabe Woodstock bought a pair of sandals in a discount store after checking carefully to be sure that they were exactly the same as a pair he had seen in a regular store.

He paid $24.98 for the sandals. In the regular store, the sandals were priced at $27.50.

$2.52

Gabe saved $_____ by buying the sandals in the discount store.

yes

Did he check quality carefully? (yes, no)

Tamra and Isabel each want to make cookies using walnuts. In the bulk section of the grocery store they see two varieties of walnuts. Each is priced per pound.

A
Bulk Walnuts $1.99

B
Chopped Walnuts $3.99

$1.99

Walnut A costs $_____ per pound.

twice

Walnut B is (twice, half) the price of walnut A.

This is because the bulk walnuts are still in the shell. The other walnuts have been shelled and chopped.

harder

The women know they need chopped walnuts for their cookies. If they buy the whole walnuts, they must shell and chop the nuts themselves. This will make the cookie-baking (easier, harder).

$3.98

Tamra decides she has the extra time to shell and chop the whole nuts. She will purchase the bulk walnuts A and save herself some money. She buys two pounds and pays a total of $_____.

$7.98

Isabel works longer hours during the week. She has the extra money, but not as much time. The chopped nuts are more convenient for her. She buys two pounds of the bulk walnuts B and spends $_____.

Each woman chose the best deal for herself and shared some delicious cookies with her family!

1/2 off means a discount of 1/2 of the regular price.

To determine what 1/2 means in pricing, we raise 1/2 to higher terms so that the denominator is 100. In this way we can write 1/2 as a decimal:

$$\frac{1}{2} \implies \frac{(1 \times 50)}{(2 \times 50)} \quad \frac{\underline{}}{100}$$

$$\frac{50}{100} \implies .\underline{}$$

Is that 50%? (yes, no)

Find the amount of discount if the regular price is $42.50 and the discount is 50%:

$42.50
x .50

amount of discount

DISCOUNT!!
50% OFF

What is the sale price?

$42.50
–_____

50 / **100**	
.50	
yes	
$21.25	
$42.50 −21.25 $21.25	

25/100	
.25	
$32.00 x .25 $ 8.00	$32.00 − 8.00 **$24.00**

SALE

1/4 OFF

$$\frac{1}{4} \implies \frac{(1 \times 25)}{(4 \times 25)} \quad \frac{\underline{}}{100}$$

$$\frac{25}{100} \implies .\underline{}$$

1/4 off means a discount of 25%.

Find the sale price of a camera bag regularly priced at $32.00, now on sale at 1/4 off. $_____ (Remember, the discount must be subtracted from the regular price to find the sale price.)

Meat is already packaged at most grocery store meat counters. Like packaged cheese, packaged meat has the cost per pound marked on it, as well as the weight of the package and the package price.

$2.69

The ground turkey costs
$_____ a pound.

1.3

The meat in the package is
_____ pounds.

GROUND TURKEY
PER POUND | WEIGHT
$2.69 | 1.3 lbs

2.69
x 1.3
$3.50

Find the package price.
(Round the answer.)
$_____

Here are two packages of a chicken roaster:

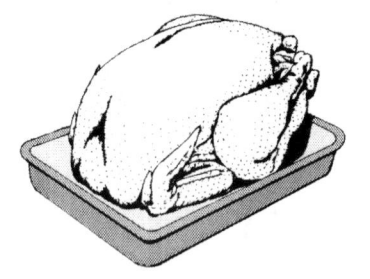

A
Cut-up Chicken $5.95

B
Whole Chicken $4.95

no

Both packages weigh the same. But does the chicken in package A cost the same per pound as the chicken in package B? (yes, no)

cut-up

Package A contains a (cut-up, whole) chicken roaster.

more

A cut-up chicken costs (more, less) per pound than a whole chicken.

You may pay more for the extra service of having the chicken cut up by the butcher. But many people do not have the time or ability to cut a whole chicken themselves, so package A could be worth the extra money.

Does a high price always mean higher quality? (yes, no)

Does a low price always mean low quality? (yes, no)

Emma bought a stove. After one month, it stopped working.

The company that made the stove fixed it. Emma did not have to pay for the repairs.

Emma bought a stove (with, without) a warranty.

60% ➠ This is a (percent, decimal).

Write it as a decimal:

60% ➠ .____

A discount is (added to, subtracted from) the regular price.

Items sell for (more, less) in discount stores.

Designer jeans that are regularly priced at $89.00 are advertised at 20% off.

20% ➠ ._____

The amount of discount is $____. ($89.00 x .20 = _____)

The sale price is $____. ($89.00 − _____ = _____)

THE END OF UNIT 3

Here are salmon fillets for $4.99 a pound.

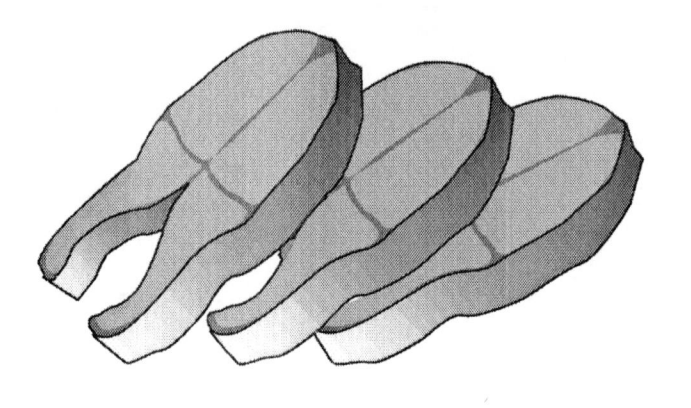

Kaitlyn is having several friends for dinner. She needs three pounds of fish. The salmon will cost her $_____.

$14.97

On another occasion, Kaitlyn decides to cook dinner for one friend and herself. Kaitlyn checks out the frozen veggie burgers. She can grill and prepare the veggie burgers in the same way as regular hamburgers.

They can eat two burgers each. There are four burgers in each package. Kaitlyn needs to buy _____ package(s) of veggie burgers.

one

The box of veggie burgers costs $3.05. A 1 lb package of ground beef can make 4 hamburgers and it costs $3.99. Does Kaitlyn (save, lose) money by buying the veggie burgers?

save

74

UNIT 10

8

smaller

wide

1

a.

When you shop for floor covering, curtains, window shades, paint, or wallpaper, you need to know **measurements.**

Nadia Klein wants a rug for her living room floor. Nadia is shopping in second-hand, or consignment stores, to see if she can find a good used rug.

Nadia's living room floor is 12 feet long and 8 feet wide. We say that the floor is 12 by ____.

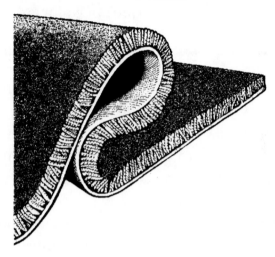

She needs a rug 12 ft by 8 ft or (larger, smaller).

She finds a pretty rug in good condition that measures 12 by 9. It will fit the floor the long way, but it is too (wide, narrow) to fit across.

Nadia decides to buy the rug and trim ____ foot off the width to make it fit. (9 – 8 = ____)

Nadia also wants a rug for her hallway. Before she goes out to buy a rug she should
      ____ a. measure the length and width of the hallway.
      ____ b. measure the height and width of the doorway.

price	You can compare foods that are priced by the pound just by looking at the (price, brand name).

Meat is priced by the pound.

Here are two kinds of steak:

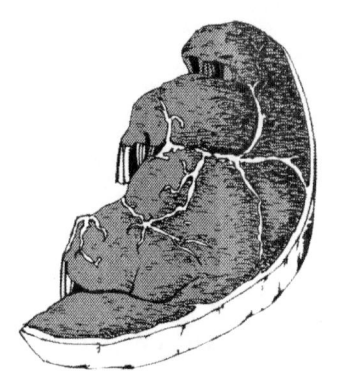 

Top Sirloin
$3.99 / lb

T-Bone
$6.59 / lb

Which kind of steak costs less per pound?
    ___ a. top sirloin steak
    ___ b. T-bone steak

a.

You may want to learn about different cuts of meat, which meats are leaner and lower in fat, and which meats are best for certain kinds of meals. A good cookbook will show you the proper handling of meat and some tasty recipes. It will give you ideas for the preparation of other foods as well.

In the produce (vegetable and fruit) department, you find fresh green beans to make for dinner. The green beans are priced at 99¢ a pound. You put green beans in a produce bag and give it to the cashier at checkout. She weighs it and tells you it is two pounds.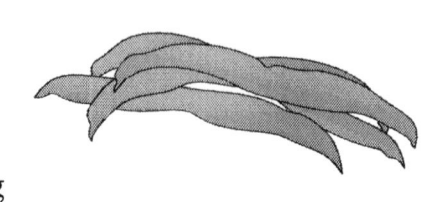

$1.98

The green beans will cost $_____.

Aden Wright wants to paint his bedroom walls light green. He must know the total amount of **wall space**, or area, in square feet so that he will know how much paint to buy.

Two of the walls are 11 by 8.

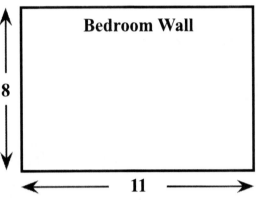

Bedroom Wall

8

11

To find the area, we multiply:

11 x 8 = _____

88

We will give the area in **square feet**. Since the area of one wall is 88 square feet, the two walls have a total area of:

88
x 2
_____
square feet

176

Here is a picture of the third wall.

Bedroom Wall

8

9

Its height is _____ feet.

8

Its base is _____ feet.

9

So the total square feet of the third wall is _____ square feet.

72

A 20 ounce jar of homestyle salsa is priced at $2.50.

Find the cost per ounce:

_____ R _____ ➠ _____ _____ ¢ per ounce

20)‾250‾   (If there is a remainder, change it to a fraction and reduce it.)

Find the cost per ounce: (Reduce fractions when necessary.)

Quantity	Price	Cost per Ounce	
9 ounces	$2.70 ➠	_____	¢
10 ounces	$2.20 ➠	_____	¢
15 ounces	$1.68 ➠	_____	¢
12 ounces	$2.46 ➠	_____	¢

To find cost per unit, we divide
   ____ a. price by quantity
   ____ b. quantity by price

Do both cans hold the same quantity of beets? (yes, no)

Beets A cost 6¢ (more, less) per pound than beets B.

A
$1.25

B
$1.31

THE END OF UNIT 9

The fourth wall has a door in it that will remain white. We can calculate how much paint is needed for this wall.

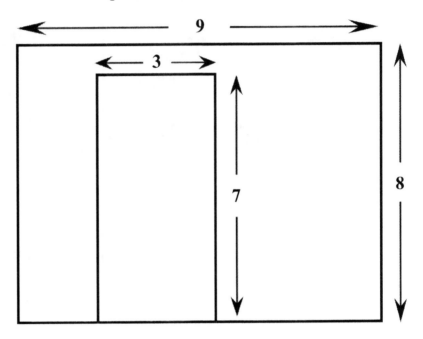

The wall is ___ feet high and ___ feet wide, so the total number of square feet in the wall including the door space is ___ x ___ = ___.

8 , 9

8 x 9 = 72

The doorway is ___ feet high and ___ feet wide, so the total number of square feet in the doorway is ___ x ___ = ___.

7, 3

7 x 3 = 21

To find out the number of square feet to be painted, we subtract the area of the doorway from the total area of the wall. ___ - ___ = ___

72 – 21 = 51

Now add the areas of all four walls:

    176
     72
   +51

        square feet

299

Aden must buy enough light green paint to cover a total of 299 square feet of wall.

pound	Tessa Sawyer is having a bake sale. She needs to buy ingredients for her special cocoa-butter cookies. To find the cost of a 1 lb package of butter, you can look at the package price.

The price on the package is also the cost per (pound, ounce). 16 (oz) ounces = 1 (lb) pound |
| **1/4, 4** | If Tessa's 1 lb package of butter contains 4 sticks, we can determine how much each stick costs.

Each stick is 1 of 4, 1/_____ of a pound, or _____ ounces.

The price on the package of butter is $3.20. |
yes	Is that the cost per pound? (yes, no)
80¢	The cost of one stick is ¼ x $3.20 = _____ ¢
	Cheese is sold by weight. The cost per pound is usually marked on the package, along with the quantity and the package price.
$4.00	This cheese costs $_____ per pound.
$2.00	The package price is $_____.
	The quantity is 1/2 lb.
yes	Is the package price correct? (yes, no) (.5 x $4.00 = _____)

AGED CHEDDAR		
PRICE PER POUND $4.00	**PACKAGE PRICE** $2.00	**WEIGHT** 8 OZ

71

99 square feet	Finally, Aden is going to paint the ceiling in his bedroom beige. What is its area? _____ square feet *(diagram: rectangle, width 9, height 11)*
3	A **pint** of paint will cover 33 square feet. Aden will need _____ pints to cover the ceiling. (99 ÷ 33 = _____)
$13.50	One pint costs $4.50, so 3 pints will cost $_____ . There are 2 pints in 1 quart, and 1 quart of the same paint costs $5.79.
true	There is as much paint in 1 quart can and 1 pint can as there are in 3 pint cans. (true, false)
yes	Can Aden save money by getting a quart and a pint can instead of getting 3 pint cans? (yes, no)
$10.29	One quart can and 1 pint can together cost $_____ .
$25.78 **$28.95**	In order to paint the walls, Aden figures he needs 5 quarts of paint or 1 gallon and 1 quart. At 19.99 a gallon (4 quarts) and $5.79 a quart, Aden will pay $_____ for paint to paint the walls. This is cheaper than buying 5 quarts of paint for _____.

UNIT 4

When you shop for packaged solid foods, you compare brands by looking at the quantities and the prices.

You can do this with any item that is sold per package and has a net weight marked on it.

A
$9.00

B
$7.00

The kitty litter in package (A, B) is less expensive.

B	

You have tried both brands. You found that you had to use twice as much of brand B as you used of brand A in order to keep the litter box clean and odor-free.

Does this mean that brand B is lower in quality? (yes, no)

yes	

We also can conclude that, if you use 1 box of brand A in a month, you would use 2 boxes of brand B in a month.

If you buy brand B, you will spend $_____ on kitty litter in one month.   (2 x $7.00 = _____)

$14.00	

If you buy brand A, you will spend only $_____.

$9.00	

So the better buy is really brand (A, B) even though it costs more per pound.

A	

By carefully comparing quality, you saved $_____.
($14.00 - $9.00 = _____)

$5.00	

Sq ft stands for **square feet.**

Aden Wright is also going to cover the inside of his garage walls with wallboard. Help him find the total area of all four walls.

70 sq ft

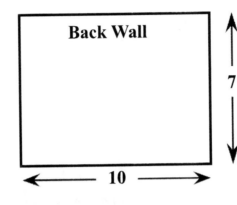

Area: _____ sq ft

84 sq ft

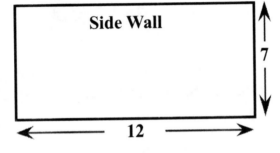

Area: _____ sq ft

12 sq ft

72 sq ft

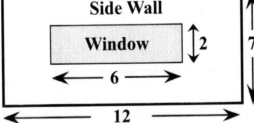

Area of *window*: _____ sq ft

Area of wall *without window*: _____ sq ft
Subtract (12 x 7) - (6 x 2).

48 sq ft

70 - 48 = 22 sq ft

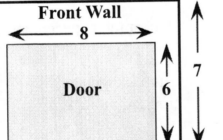

Area of *door*: _____ sq ft

Area of wall *without door*: _____ sq ft

Total area: 70 sq ft
84 sq ft
72 sq ft
+22 sq ft
_____ sq ft

248 sq ft

If the price is the same for two packages, but the quantity in one is larger than the quantity in the other, you know that the larger quantity costs

A
$2.49

WHOLE WHEAT
SODA CRACKERS
NET WT 1 lb 5 oz

B
$2.49

SODA
CRACKERS
NET WT 1 lb 8 oz

b.

_____ a. more per unit.
_____ b. less per unit.

B

Which package holds more crackers by weight? (A, B)

less

So the crackers in package B costs (more, less) per ounce.

$$\overset{12¢}{12\overline{)\$1.44}} \qquad \overset{14¢}{12\overline{)\$1.68}}$$

quantity

The number we divide by is the (price, quantity).

larger

If the quantity stays the same, but the price gets larger, the cost per unit also gets (larger, smaller).

COCOA
1 lb

A
$1.44

B
$1.68

COCOA
16 oz

Look carefully at the quantities and prices of these two cans of hot cocoa.

A

The hot cocoa in can (A, B) costs less per pound.

69

UNIT 9

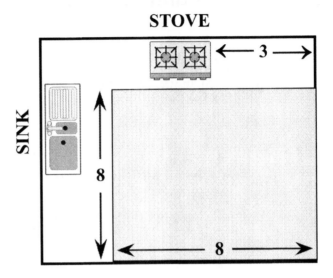

STOVE

SINK

The above picture is the floor plan for Nadia Klein's kitchen.
Nadia wants to cover the floor (the shaded part) with ceramic tiles.

We will calculate how many square feet this is.

The floor is made up of 2 squares. One is 8 x 8 and the other is 3 x 3.

64

There are _____ square feet in the 8 x 8 square.

9

There are _____ square feet in the 3 x 3 square.

73

That makes a total of _____ square feet.

Ceramic tiles cost $1.51 a square foot. So Nadia will pay:

$1.51
x 73

$110.23

30 1/2¢	Find the cost per ounce: (Reduce the fractions.)

Find the cost per ounce: (Reduce the fractions.)

Quantity	Price	Cost per Ounce
8 ounces	$2.44	_____ ¢
12 ounces	$1.58	_____ ¢
1 pound	$2.60	_____ ¢

30 1/2¢

13 1/6¢

16 1/4¢

a.

To find the cost per unit, we divide
_____ a. price by quantity.
_____ b. quantity by price.

You do not always have to calculate the exact unit price of two items in order to determine which item is the better buy. Sometimes the better buy is easy to see.

For example, which is the better buy?
_____ a. a 10 oz bag of potato chips for $2.80
_____ b. a 16 oz bag of potato chips for $3.00

b.

You can see that the 16 oz bag weighs about 1½ times as much as the 10 oz bag but costs less than 1½ times as much.

a.

Which is the better buy?
_____ a. 1 lb box of brown sugar for 95¢
_____ b. 14 oz box of brown sugar for $1.05

b.

Which is the better buy?
_____ a. a 12 oz box of oatmeal for $3.66
_____ b. a 1 lb box of oatmeal for $3.80

Aden decides to paint his recreation room navy blue. The room has a door along one wall that will not be painted and a window along another that will not be painted.

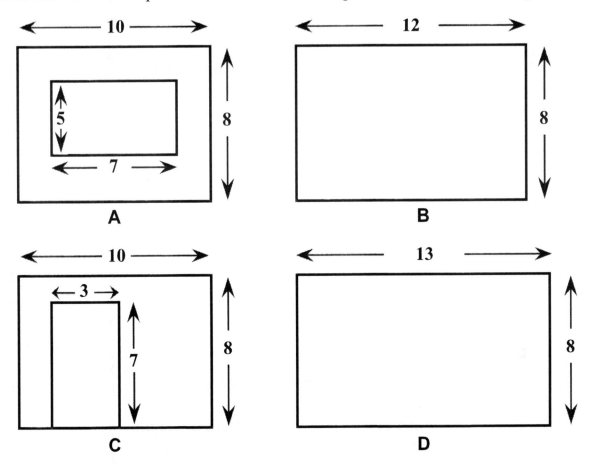

These pictures show the length and width of each wall, of the window, and of the doorway.

The total area of A including the window is ___ x ___ = ___.

The total area of the window is ___ x ___ = ___.

The area of A to be painted is ___ - ___ = ___.

The area of B to be painted is ___ x ___ = ___.

The total area of C including the doorway is ___ x ___ = ___.

The total area of the doorway is ___ x ___ = ___.

The total area of C to be painted is ___ - ___ = ___.

The area of D to be painted is ___ x ___ = ___.

The total area in the recreation room to be painted is ___ + ___ + ___ + ___ = ____.

THE END OF UNIT 4

larger

Often you can get a better deal by buying a (smaller, larger) quantity.

Good Morning CEREAL — NET WT 18 OUNCES — $2.88

Good Morning CEREAL — NET WT 12 OZ — $2.22

Chou Lee is doing his weekly shopping. On the cereal shelf he finds two boxes of the same kind of corn flakes.

$2.22

The 12-ounce box is priced at $_____.

Use the space below to find the cost per ounce. (Divide. Write the remainder as the numerator of a fraction. Then reduce the fraction.)

18 1/2¢

Cost per ounce: ____ ____¢

$2.88

The 18-ounce box is priced at $_____.

16¢

That is _____¢ per ounce.

larger

Chou pays less per ounce by buying the (larger, smaller) box.

smaller

But if Chou cannot eat all 18 ounces of corn flakes before they get stale, it would be wise for him to buy the (larger, smaller) box.

Many people own a car at some time in their lives. For some people, a car is the most expensive property they own.

Because buying and maintaining a car is so expensive, it is important to understand how to buy wisely and protect your investment.

The first question to ask when buying a car is, "What do I want the car for?"

If you are 17 and do not have much money, which car makes more sense to buy?

_____ a. a used, less-expensive car
_____ b. a new ($20,000) car

If you have a family of four, which is probably the better car for you?

_____ a. a subcompact convertible (seats 2)
_____ b. a sports utility vehicle (seats 7)

If you are 65 and can afford an expensive car, which car might you prefer?

_____ a. a comfortable luxury car
_____ b. an Italian, 2-door sports coupe

If you travel a lot for business purposes, one feature you would want in a car is

_____ a. good gas mileage.
_____ b. an all-leather interior.

a.

b.

a.

a.

$0.09	When we write 9 cents as a decimal, we have to put a 0 in the tenths place. 9 cents ➞ $.0_____
do not need	But when we write it with the cents sign, we (need, do not need) the 0.
9¢	$0.09 ➞ _____¢
5¢	$0.05 ➞ _____¢

$4.30	When we have more than 100 cents, we rewrite the amount in dollars. 430¢ ➞ $_____._____
	A 12-ounce box of whole grain cereal is priced at $2.46. Find the cost per ounce:
20 R6	$$\begin{array}{r} \text{R6} \\ 12\overline{)\,246} \\ -240 \\ \hline 06 \\ -0 \\ \hline 6 \end{array}$$
	We can write the remainder as the numerator of a fraction. The denominator will be the number we divided by. Reduce: $\dfrac{6}{12} \div \dfrac{6}{6} = \dfrac{}{2}$
1/2	
20 1/2¢	So the cost per ounce of the cereal is 20_____¢
a cent	When we are dividing cents to find cost per unit, any remainder is a fraction of (an ounce, a cent).

There are two different ways to go about buying a car.

You can buy a new car. Which of the
following are true about new cars?

a. _____ a. They are in top condition because they have never
          had an owner.
       _____ b. They use more gasoline per mile than used cars.
c. _____ c. They cost more because they have never had
          an owner.

You might prefer to buy a used car. Which of the following are true
about used cars?

a. _____ a. They have had at least one owner.
       _____ b. Like wine, they improve with age.
c. _____ c. They break down more often than new cars because
          they are older.
d. _____ d. The older cars get, the cheaper they get.

A person might choose a new car over a used car because

a. _____ a. it is hard to be certain that the used car does not have
          some hidden problems.
       _____ b. new cars generally are cheaper

Net weight is

    ___ a. the weight of the package.

    ___ b. the weight of the food inside the package.

Weight is given in

    ___ a. ounces and pounds.

    ___ b. feet and inches.

Does the size of a package always show the quantity of food inside? (yes, no)

The weight of nuts in these cans is (the same, different).

Look at the package prices.

(Peanuts, Cashews) are more expensive.

$1.95

$3.59

A
$3.60

B
$3.20

Bag A holds ___ ounces of potato chips.

The cost of the chips in bag A is ___¢ per ounce.

Bag B holds ___ ounces of potato chips.

The cost of the chips in bag B is ___¢ per ounce.

The chips in bag (A, B) are more expensive.

THE END OF UNIT 8

If you need a car, you do not always have to buy one.

Many people **rent** cars. You can rent a car for a day, 3 days, a week, or however long you like. Generally you pay a fixed rate per day (or week), and then pay an extra amount depending on the number of miles you drive.

You might rent a car if (more than one answer is possible)

a.   _____ a. you do not often need a car, and yet you want one for a special occasion.

b.   _____ b. you have flown to another city and need a car while you are there.

c.   _____ c. your regular car is being repaired and you need transportation.

You can also lease a car. Usually, leases for a car start at 24 months. You pay part of the lease cost each month until the end of the lease period.

renting

Leasing a car is like (renting, buying) a car.

Think carefully when deciding between renting or leasing a car. When you rent a car, you can return it when you feel like it. When you lease a car, it is yours until the contract ends. Trading-in or exchanging vehicles is not allowed.

less

If you leased a car for a year, it would probably cost (more, less) than if you rented the same car for the same period.

A
$1.20

B
$0.85

1	Can A holds _____ pound of baked beans.
8	Can B holds _____ ounces of baked beans.
2	There are 16 ounces in a pound, so to get a pound of baked beans in smaller size cans, you would have to buy _____ cans. (16 ÷ 8 = _____)
$1.70	The price of can A is $1.20. The price of can B is 85¢. 2 cans of brand B would cost _____.
a.	If you want to buy a pound of baked beans, you would save money by getting _____ a. 1 can of Homestyle Baked Beans. _____ b. 2 cans of Good to Eat Beans.
	Some foods, like bread and tortilla chips, have the net weight printed on the package.
quantity	This is so the buyer can compare (quantity, quality).
cost	In a grocery store, prices are displayed on labels near or on food items. This is so that the buyer can see the (cost, quality).
	The best way to know the quality of food is to try it. Nutrition facts are listed on each item. Read the list of ingredients on the package. This tells you what was used to make the food. Ingredients are listed in order of the percentage that composes the product. The main ingredient is listed first. The one that makes up the smallest percentage of the product is listed last.
quality	The list of ingredients can help you to compare (quantity, quality).

Suppose you are interested in buying a new car. What do you do?

First, it is a good idea to know whether you want the car for long distance driving, hauling materials, off-road driving, or short trips around town. You should also know what price you are willing to pay.

You should have an idea what the car will be used for before you start to look because
> _____ a. it never hurts to be narrow-minded.
> _____ b. when you know what you want, you are less likely to purchase an item you do not want.

b.

Where do you look first? It is a good idea to start shopping at a known dealer or a dealer recommended by a friend. Do not feel you have to buy your car at that place. In the beginning, you are trying to figure out what you want–what make and model of car matches your needs and what optional features interest you.

The make of a car refers to the manufacturer: Toyota, Ford, Pontiac, or Honda, for example.

The model of a car refers to the particular design, such as Toyota Camry or Ford Mustang.

When you are looking around at cars at a dealership,
> _____ a. do not ask too many questions of the salesperson because then you may feel obligated to buy a car.
> _____ b. ask any questions to get a better idea what you want. You have no obligation to the salesperson.

b.

A
$1.95

B
$2.15

Here are two brands of chocolate chip cookies.

Compare the quantities. (There are 16 ounces in 1 pound.)

the same

The quantities are (the same, different).

can

Since the quantities are the same, we (can, cannot) simply compare the package prices.

B

Brand (A, B) is more expensive.

A

If you like both brands of cookies, your best buy is probably brand (A, B).

You can buy a 1-pound can of crushed tomatoes for $1.35.

You can buy two 1-pound cans for:
$1.35
x   2

$2.70

A 2-pound can of the same brand of crushed tomatoes is priced at $2.60.

If you make a lot of tomato sauce for spaghetti, your best buy is
_____ a. two 1-pound cans.
_____ b. one 2-pound can.

b.

When you are looking at cars in a lot, you will see a price sticker on each car. The sticker totals up all the expenses you will have to pay for the car.

Keep in mind, many cars now come equipped with certain standard items, such as an AM-FM stereo CD player, anti-lock brakes, and air-conditioning.

The sticker below lists optional equipment for this particular car. They are not necessary add-ons, but can be thought of as luxury items.

Consider one such sticker:
Manufacturer's Suggested Retail Price (MSRP): $17,900.00
Optional Equipment (Add-ons):

6-speed transmission, automatic	1,075.00
Rear spoiler	479.00
Rear passenger side airbags	350.00
Rubber mats (set of 4) andTrunk liner	185.00
Cargo net	69.00
Destination fee	630.00
Total MSRP:	$20,688.00
Sales tax (6%):	1241.28
License and registration:	120.00
Total price:	$22,049.28

The manufacturer's suggested retail price (MSRP) is

_____ a. the total amount you pay.

b. _____ b. the base price of the car, without the extras, sales tax, and the license and registration fees.

The total MSRP is

a. _____ a. the total amount of the car including the optional equipment and destination fee.

_____ b. the final price of the car after tax has been included.

Good Morning CEREAL
NET WT 18 OUNCES
A
$3.60

GRAIN O'S CEREAL
NET WT 15 OUNCES
B
$3.00

Suppose you have tasted both cereals and equally like them. You read the ingredients; they are the same too!

To you, both cereals are of the same quality.

Since you like both cereals just the same, and the cost per ounce is the same, you will decide which box to buy according to how much cereal you want.

quantity	How much is (quantity, quality).
larger	If you eat a lot of cereal, you will probably buy the (larger, smaller) quantity of cereal.
A	You will buy box (A, B).
8	This box holds _____ ounces of crackers.
quantity	That is the (quantity, quality) of crackers.
$3.20	The package price is $_____.
40¢	The cost per ounce is _____¢.

WHOLE WHEAT CRACKERS
8 OZ ($3.20)

a.

The sticker shown on the previous page is attached to a certain car on the lot. The sticker is letting a person know that the car

_____ a. has a 6-speed automatic transmission and rear passenger side airbags, among other features.

_____ b. is a floor model not available for sale

b.

There is a destination fee listed on the sticker below the add-ons. This is what the car manufacturer charges for the delivery of your car from the production facility to the dealership.

In the state in which the car is being sold, there is a 6% sales tax on

_____ a. the MSRP (Manufacturer's Suggested Retail Price)

_____ b. the total MSRP.

a.

All states require a new car to have a license plate. A license plate is used

_____ a. to clearly identify each car on the road.

_____ b. to give the front or back of your car extra class.

e.

New, unpurchased cars do not actually have license plates. In many states when you buy the car, you pay for a vehicle license and registration as part of the total cost. The dealer then puts a temporary license on the car. It is a rectangular piece of paper, with large colored numbers on it. This allows you to drive your car on public roads until you receive your actual metal license plates to put on.

The total price of the car is

_____ a. the suggested retail price.

_____ b. the price of the options.

_____ c. the destination fee.

_____ d. the sales tax.

_____ e. the sum of all the above.

You compared the quantities of cereal in these two boxes.

You found that box (A, B) held more.

A

A
$3.60

B
$3.00

Now you should compare cost.

Since the boxes hold different quantities, you (can/cannot) compare the package prices.

cannot

The cereal in box A costs _____¢ per ounce.

$(360 \div 18 =$ _____ $)$

20¢

We are using 1 (ounce, pound) as our unit.

ounce

The cereal in box B costs _____¢ per ounce.

20¢

To find the cost per unit, we divided the price by the quantity.

The cost per ounce of the cereal in box A is the same as the cost per ounce of the cereal in box B.

Is one brand of cereal more expensive than the other? (yes, no)

no

Suppose you wanted to buy the same sort of car as the one with the sticker on it. However, you do not want any of the options. If you were to buy the car from the dealer, he would

_____ a. take all the extras out of the car you are looking at and sell your car in a stripped-down form.

_____ b. see if he had a car in inventory (on the lot) with what you wanted on it. If not, he would have to send for a car from the factory that matched your exact order.

Suppose you wanted the same model as the car with the sticker on it, but you only wanted one option, the rear passenger side airbags.

Calculate how much it would cost.

b.

The MSRP is $_____ $17,900.00

The airbags are $_____ $350.00

The destination fee is $_____ $630.00

This makes a total MSRP of $_____ $18,880.00

The sales tax is:
$ _____ ← total MSRP
x ____.06 ← sales tax

$18,880.00
x .06
$1,132.80

The license and registration are $_____. $120.00

The total price is $_____. $20,132.80

Dry foods that are sold in packages show quantity as net weight. This is the weight of the food only, not the weight of the box or can.

Using the U.S. Customary Weights and Measures System, weight is given in **pounds** and **ounces**.

Net weight (wt) is the weight of
____ a. the cereal only.
____ b. the box and the cereal.

The cereal in box A weighs _____ ounces.

The cereal in box B weighs _____ ounces.

Are the boxes the same size? (yes, no)

The size of a package does not always show the quantity of food inside.

The quantity of cereal in the boxes above is shown
____ a. as net weight in ounces.
____ b. by the sizes of the boxes.

You would get more cereal by buying box (A, B).

a.

18

15

yes

a.

A

Before you start looking for an automobile, it is good to have a basic idea of what you want the car for and what you are willing to pay. (true, false)

Which of the following are true?

_____ a. New cars are in better mechanical condition than old cars.
_____ b. Generally speaking, the older the car the higher its value.
_____ c. It is cheaper to rent a car for a year than to lease it for a year.
_____ d. If you take up the salesperson's time, it is understood that you will buy a car from him.

The following is part of a sticker on a car window:

Manufacturer's suggested retail price (MSRP):	$20,475.00
Removable roof rack:	375.00
Fog lights:	330.00
Destination fee:	595.00

The cost of the basic car with these three items is $_____ .

Sales tax is 5% so the tax will be $_____ .

A vehicle license and registration is $130.00.
The total price of the car including tax,
and registration is $_____ .

THE END OF UNIT 5

Fl oz stands for (quart, fluid ounce).

1 qt = _____ fl oz

Milk (A, B) is more expensive.

It costs _____¢ more per quart.

$1.36 $1.39

When we want to compare the cost of two packages holding the same quantity, can we simply look at the package price? (yes, no)

Bottle A holds _____ fl oz of liquid hand soap.

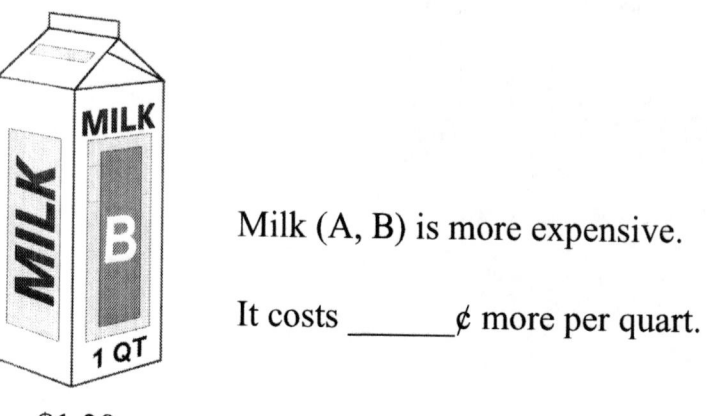

A
$0.88

B
$1.34

Cost per fl oz: 8)‾88¢‾

12 fl oz of hand soap A would cost _____¢.   (12 x 11¢ = _____)

12 fl oz of hand soap B costs _____¢.

Liquid hand soap (A, B) is more expensive.

You find a car you like. It seems to fit all your requirements. The price is affordable. The style is sharp. At this point you are ready to test-drive the car to see how it feels.

Any reputable dealer will let you drive the car. You are under no obligation to buy the car. Some dealers will let you drive the car alone; others will want the salesperson to drive with you.

CAUTION: Before you drive the car, be sure the dealer has insurance on the car because

a.

_____ a. in case you get in an accident, you do not want to have to pay for the damages.

_____ b. that is the best way to tell if you are getting a good deal on the car.

After you have decided what model of car you want and what features suit your needs best, it is time to find the place where you will buy the car.

Look around! Different dealers will start out by quoting the manufacturer's suggested retail price, which is a selling price figure supplied by the factory. In fact, however, most dealers will finally sell you the car at a lower price because

a.

_____ a. bargaining for a better deal is expected to happen between the customer and the salesperson.

_____ b. in today's economy, nobody can afford the high price of automobiles.

When we divide price by quantity, we are finding the cost per unit. 12 fl oz of imitation maple syrup is priced at 84¢.

Each fluid ounce is a unit. Find the cost per unit:

7¢

$$12\overline{)\,84¢\,}$$

A quart, 32 fl oz, of the same syrup at the same cost per unit, would be priced at:

$$
\begin{array}{r}
32 \\
\times\ 7¢ \\
\hline
224¢
\end{array}
$$

more

224¢ is (more, less) than a dollar.

So we write it as dollars and cents. We need 2 decimal places:

$2.24

224¢ ⟹ $2._____

fluid ounce

2 fl oz of liquid hand sanitizer is priced at $1.20.
The unit we are using is the (fluid ounce, pint).

Find the cost per unit:

60¢

$$2\overline{)\,\$1.20\,}$$

If you presently have a car that you are willing to replace, you can offer the car as a trade-in on the new car.

If you trade in your old car on the new one, that means

a.

_____ a. the dealer will keep your old car and subtract its current value from the price of the new one.

_____ b. the dealer is willing to make an even trade with you — your old car for his new one.

You can find out the approximate value of your old car by looking at the Kelley Blue Book. The Blue Book is a dependable resource used by anyone looking for new car prices, used car values, rebates (discounts given to purchasers) and more.

You can easily calculate the value for your car by going online to the Kelley Blue Book's resource website, www.kbb.com

Many dealers will quote you the blue book value of your car over the phone. However, this is a estimate. To find out exactly how much your car is worth, bring the vehicle directly to the dealer.

You can also get an idea of what your old car is worth in your area by (more than one answer is possible)

a.

_____ a. looking at want ads in the newspapers for cars similar to yours in model and year.

_____ b. determining what the original cost of the car was.

c.

_____ c. looking at the price on cars similar to yours in model and year in used car lots.

Here are two brands of bottled water.

Bottle (A, B) holds the greater quantity and has the higher price.

A $0.72

B $0.96

We cannot assume that brand B is more expensive until we compare the cost per unit.

In this case, our unit will be the
_____ a. fluid ounce.
_____ b. quart.

Brand A:

_____ ¢ per fluid ounce

12)‾7‾2‾¢‾

So 16 fl oz of brand A would cost:
16
x 6¢
‾‾‾‾

By comparing the cost per unit, we find that 16 fl oz of brand A costs the same as 16 fl oz of brand B. The two brands cost the same for the same quantity.

You will decide which one to buy depending on how thirsty you are, and, if you have tried both brands, which one you like better.

If you think both brands are of the same quality, you have only to decide
_____ a. which one is cheaper.
_____ b. how much you need.

B

a.

6¢

96¢

b.

Once you have decided which car you want, you can find out what trade-in allowance the dealer will give for your old car (if you plan to trade one in). The difference is what the new car will cost.

At this point, in order to get the best value for your money (more than one answer is possible),

_____ a. Visit some other dealers. Perhaps they will put a higher trade-in value on your old car.

_____ b. Consider selling your old car by yourself. You might get more money from a private individual than you can get from a dealer.

_____ c. Wait a year. Next year's model will be cheaper.

_____ d. Go to other dealers. Tell them what your dealer is offering. Another dealer might make you a better offer.

The total price of the car as shown on the sticker is rarely the price you have to pay, provided you are careful.

The dealer wants to make the sale. She does not want you to go to one of her competitors.

If you are really interested in buying the car, bargain with the dealer! Make an offer lower than her price.

You should make an offer only if you are serious about buying the car. (true, false)

The dealer now knows you are interested. She may now make a counter-offer.

When you bargain with the dealer, you discover what the dealer really thinks her car is worth. (true, false)

a.

b.

d.

true

true

can	Each container holds the same quantity of apple juice. So we (can, cannot) simply compare the package price.
	 $2.69 $2.89 
B	Apple juice (A, B) is more expensive.

---

no	Can we only compare the package prices of these two cartons of orange juice? (yes, no)
	 $1.60 $2.80 
different	This is because the quantities are (the same, different).
$1.60	We will use the pint as our unit. 1 pt of orange juice A costs $_____.
$1.40	1 pt of orange juice B costs $_____. ($2.80 ÷ 2 = _____)
A	Orange juice (A, B) costs more per pint.
	This may be simply because it is packaged in a smaller quantity. Usually, the smaller packaged quantity, the higher the cost per unit.
yes	If you need a large quantity, can you sometimes save money by buying a larger package? (yes, no)

If you and the dealer can agree on a price, you can probably make a deal.

You may not agree on a price, though. Suppose that the total sticker price of the car is $22,000. You offer $18,000. The dealer offers $21,000. You offer $20,000. The dealer refuses to go lower.

What can you do? If you go to $21,000, you will have a deal. Remember, though–you might be able to get the car for $20,000 at some other dealer.

If you are not happy with the price the dealer offers, you can always

b.

_____ a. complain to the Better Business Bureau.
_____ b. find out what other dealers will charge for the same model of car.

Now, suppose that you find another dealer who asks $20,500 for the car. Should you accept his offer over the offer of the dealer asking $21,000? Not necessarily!

There are some other factors to consider, besides the price, that may make the $21,000 deal a much better one. Does the dealer offer a good warranty on the car? Will the dealer answer all your questions about the car's performance?

true

The responsibility of owning a new car does not end on the day you make the purchase. (true, false)

It would be hard to compare the price per bottle, or package price, of two kinds of vinegar, because one bottle holds more than the other. In most grocery stores, therefore, the store owner puts a shelf sticker near different items, which gives us the cost per unit of weight. This helps us to choose more for our money.

We know that 100 cents equals 1 dollar. If we have an item that costs more than 100 cents, we can find the dollar price. An item costs 140¢, or $1.40.

We can also change dollars to cents by removing the decimal.

240¢

$2.40 ➡ _____ ¢

Then we can easily divide the cost by the **ounces** to find the cost per unit.

30¢

$$8\overline{)240¢}$$

So, price per unit is then calculated like this:

30¢

Balsamic vinegar: $$\underline{\hspace{2cm}}¢ \text{ per fluid ounce}$$
$$8\overline{)\$2.40} \text{ (or } 240¢)$$

20¢

White wine vinegar: $$\underline{\hspace{2cm}}¢ \text{ per fluid ounce}$$
$$16\overline{)\$3.20} \text{ (or } 320¢)$$

We do not need to add the decimal places because we are expressing the amount in cents.

balsamic

The (white wine, balsamic) vinegar costs more per fluid ounce. We can infer that balsamic vinegar is more expensive than white wine vinegar.

cannot

When we want to compare the cost of two packages holding different quantities, we (can/cannot) simply compare the package price.

We have to compare the cost per unit.
The unit we used above in comparing two kinds of vinegar was the

a.

____ a. fluid ounce.
____ b. square foot.

55

UNIT 7

How about $300 a month?

For 4 years?

When buying a car, the most common practice is to work out an arrangement of monthly payments that lasts until the car is completely paid for.

Suppose you are planning to buy a new car at a total cost of $21,000. The dealer allows you a trade-in value of $6,000 on your old car.

$15,000

Subtracting the trade-in value of the old car, you owe $_____.

You will not have to pay the full amount of your car right away. You will pay for the car in monthly installments over several years.

The word installment means
_____ a. repair
_____ b. payment

b.

You will then finance your car through the dealer. That means you can have the car now, but you have time to earn enough money to pay for it.

The dealership will have a captive finance company loan you the money, not the actual person from whom you purchased your vehicle. A captive finance company is usually owned by one of the big automotive companies. For example, Ford Motor Credit Company will loan money to Ford customers so they can finance a car.

true

If you finance your car through the dealership, you will probably be making your payments to a finance company. (true, false)

16 fl oz	Liquid measurements are often given in fluid ounces. Fl oz stands for fluid ounces. 32 fl oz = 1 qt 2 pt = 1 qt _____ fl oz = 1 pt
16 fl oz	The quantity of salad dressing in this bottle is _____ fl oz. SALAD OIL 16 fl oz
yes	Is that 1 pint? (yes, no)
8 fl oz	When grocery shopping, you easily can compare price on the same quantities. It is a bit harder to compare quality. Part of what you decide about quality depends on what you like. Even though these dressings are used in similar ways, one is labeled organic and the other one is not. Organic OLIVE OIL 8 fl oz $7.95 $5.95 CORN OIL 8 fl oz Both containers hold _____ fl oz of salad dressing.
	Organic means a food is grown without the farmer using pesticides, antibiotics, or growth hormones. Organic products, such as dressings, contain ingredients that are grown this way.
more	However, organic foods cost (more, less) than foods that are not organic.
	Some people prefer to buy organic products because they feel better knowing how the food items are grown.
save	Buying organic foods is not affordable for many people. If you prefer the taste of a cheaper brand over an organic one, then you may decide to buy it and (save, lose) money.

$11,250	In order to finance the car, you will need to make a down payment.
	The dealer offers you the following payment plan: Make a **down payment** of $3,750 now, and pay $245 a month for 60 months.
	If you now pay $3,750 of the $15,000, how much will you still owe? $_____
	The dealer lets you pay off the $11,250 over a period of time, but he will charge you extra money for this arrangement. This charge is called a finance charge.
b.	He charges you this extra money because _____ a. at the end of 60 months the price of a new car will be higher. _____ b. he has to wait for his money instead of getting it all right now.
	We can calculate how much this finance charge is.
$14,700	If you pay $245 a month for 60 months, how much is your total payment? $_____
$3,450	Suppose that you plan to pay off $11,250 in monthly installments. The total amount you pay is $14,700. That means you are paying a finance charge of $_____ in order to pay for your car in this manner.
5	60 months is _____ years.
$690	$3,450 extra in 5 years is an additional $_____ each year.

Compare the cost per quart:

$2.96 $3.05

Chocolate soy milk (A, B) costs more per quart.

Cream is sold in half-pints as well as in pints.

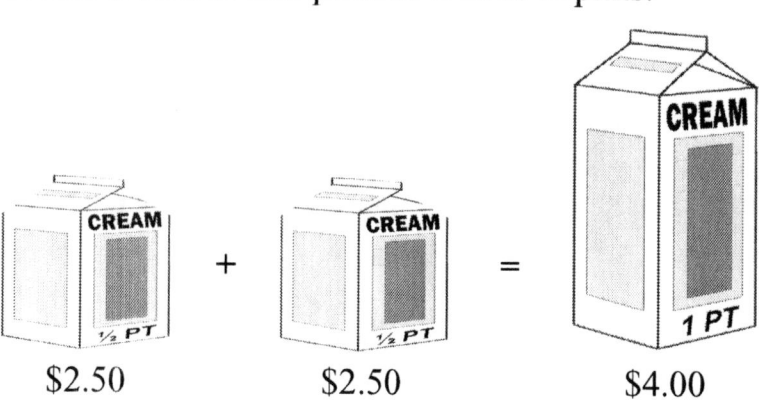

$2.50 $2.50 $4.00

2 half-pints of this cream cost $_____. (2 x $2.50 = _____)

1 pint costs $_____.

If you use a lot of cream, you could save money by buying
    ___ a. 2 half-pints.
    ___ b. 1 pint.

In other words, if you can use a large quantity, you can save a little money by buying the larger package.

Keep in mind, if you do not use much cream, you should probably buy 1 (half-pint, pint) because cream will spoil if it is kept too long.

B

$5.00

$4.00

b.

half-pint

After looking over the dealer's installment plan, you may decide you can get better financing elsewhere.

Better financing means
_____ a. your finance charge will be lower.
_____ b. your financing comes from a better person.

a.

You also can get financing from a local bank. If you belong to a credit union, you can check there for financing.
You are free to find the installment plan that suits you best.
 (true, false)

true

Installment plans for large amounts are often drawn up for 60 months (or 5 years), but smaller amounts can be financed for just 3 years.

Perhaps your dealer has a plan in which you pay the $11,250 off in 3 years. The finance charge will not be as high for a 3-year plan as it would for a 5-year plan because
_____ a. you will have paid twenty-four months less interest.
_____ b. the price of new cars 3 years from now will be cheaper than the price 5 years from now.

a.

Whole milk has a certain amount of cream or fat in it, at least 3.25%.

2 % milk, reduced fat, has 2% cream in the milk.

1% milk, lowfat, contains 1% cream.

Skim milk is milk in which the cream has been removed.

Soy milk is a dairy free product that people drink in place of milk, usually for health reasons.

Which has more cream?

_____ **a**. 2% milk

_____ **b**. skim milk

a

Milk with cream in it usually costs more than milk with less or no cream.

Whole milk costs (more, less) than skim milk.

more

Soy milk can cost more than any kind of dairy milk. However, it is low in fat and a tasty substitute if you have allergies to dairy products. Can the higher priced soy milk be a wise choice?

yes

Milk is a liquid. It is sold in containers that hold a standard measure.

Standard means always the same.

The quantity of milk in a 1-quart bottle or carton, no matter what brand of milk it is, is always _____ quart.

1

When you compare the prices on two quarts of soy milk of different brands, you are comparing the cost per quart.

When you compare the prices on two pints of soy milk, you are comparing the cost per (quart, pint).

pint

52

$21,000
x   .30
$6,300

$3,150

$14,700

$14,700	$14,700
x .15	- 2,205
$2,205	**$12,495**

a.

a.

b.

Used cars are much cheaper than new cars. The value of a new car depreciates 20% to 30% the first year and 15% to 20% each year.

To depreciate means to drop in value.

Suppose a new car costs $21,000. 30% of $21,000 is $_____.

15% of $21,000 is $_____.

If a $21,000 car depreciates 30% the first year, it is worth $_____ at the end of that year.

If it depreciates another 15% the second year, it is worth $_____ at the end of the second year.

Used cars can be good deals. It is very important to be careful about the mechanical condition of the car. A used car may be shiny on the outside,

_____ a. but the car could break down two months after you buy it.

_____ b. which means that it is as good as a brand new car.

If you are planning to buy a used car, you may need to (you may check more than one)

_____ a. have a mechanic check the car.

_____ b. take the car to a state inspection center to have it checked.

_____ c. accept the dealer's (owner's) word that the car is in good shape.

When we **compare** items, we look at them and decide which is better.

 99¢   97¢

We compared the quantities of 2% milk in these two cartons. Both cartons hold the same quantity.

yes

Did we also compare the prices of the cartons? (yes, no) One carton costs more than the other.

Different brands of 2% milk may taste different to you. This may give you an idea about the quality of the milk.

B

If the quality of milk A is about the same as the quality of milk B, milk (A, B) is the better buy because it costs a little less.

A

If the quality of milk A is better than that of milk B, you may feel it is worth the extra 2¢ to buy milk (A, B.)

b.

If you buy milk A because you feel it is better, you will be paying extra for

_____ a. more quantity.
_____ b. better quality.

Which two of the following are true?

_____ a. The sticker price on a car is the price you must pay.

_____ b. Always be sure the dealer has insurance on a car you are test-driving.

_____ c. The Kelley Blue Book is a resource used to determine the value of a car.

_____ d. An installment is the building in which horse stalls are found.

Suppose you are going to buy a car costing a total of $22,000.
You make a down payment of $5,500. How much more do you owe? $_____.

The dealer offers you a plan that allows you to pay off the amount by paying $365.70 a month for 60 months. In all, you will pay $_____ in installment payments.

That means you are paying $_____ as a finance charge on your installment plan.

Suppose you buy a new car for $18,000. If the car depreciates 30% the first year, how much is it worth at the end of the first year? $_____

If the car depreciates 15% more the next year, how much is the car worth at the end of the second year? $_____
(Hint: $12,600 x 15% = _____)

THE END OF UNIT 6

When packaged foods are priced at less than $1.00, grocery stores sometimes advertise these items using a cent sign (¢), instead of a dollar sign and decimal point.

$$9¢ \quad = \quad \$0.09$$

$0.18

$$18¢ \quad = \quad \$.\underline{\qquad}$$

When we write 9 cents as a decimal, we place a 0 in the tenths place.

However, when we use the cent sign, we do not need the 0.

$$\$0.09 \implies 9¢$$

5¢

$$\$0.05 \implies \underline{\quad}¢$$

A grocery store sells two brands of 2% reduced fat milk, brand A and brand B.

99¢

97¢

1

The quantity of 2% milk in each carton is ___ pint.

99¢

1 pint of 2% milk A costs ___¢.

97¢

1 pint of 2% milk B costs ___¢.

---

more

2% milk A costs (more, less) per pint than 2% milk B.

You buy an item priced at $19.79. You give the cashier a 20-dollar bill and 4 cents, or a total of $_____ .

The cashier owes you 2 dimes. Is that the right change? (yes, no)

The cashier still owes you _____ cent(s).

A dress regularly priced at $36.00 is on sale at a discount of 10%.

10% ➡   ._____

The amount of discount is $_____ .

The sale price is $_____ .

A wall is 5 ft by 8 ft. What is its area? _____ sq ft

If a pint of paint covers 20 square feet, how many pints are needed to paint the wall? _____

If a pint of paint costs $3.09, how much will it cost to paint the wall? $_____

When you buy a new car, it is a good idea to (you may check more than one)
        _____ a. get prices from several dealers first.
        _____ b. accept the sticker price of the car as final.
        _____ c. know how much finance charge you are paying the dealer if he
                finances your car.

THE END OF SIDE 1

48

Just as you shop for the best buys in clothing, appliances, furniture, and other goods, you also shop for the best buys in groceries.

Many of the foods you buy are sold in packages (boxes, jars, cans, or cartons).

In the U.S. Customary Weights and Measures System, liquid quantities are shown as fluid ounces, pints, quarts, and gallons. (fl oz, pt, qt, gal)

In the International System of Units, liquid quantities are measured in milliliters, centiliters, and liters. This measurement standard is known as the metric system. (ml, cl, l)

The U.S. commercial industry does not often use the metric system for calculating the weights and measures of consumer-purchased products.

Qt stands for quart.

The quantity of milk in this carton is _____ quart.

Pt stands for pint.

2 pints = 1 quart

L stands for liter.

1 quart = .95 liter (rounded)

If we poured the milk from these two 1-pint cartons into a pitcher, what total quantity of milk would we have in the pitcher?
(1 qt, 1 pt)

1 qt = ___ pt

2 qt = ___ pt

1

1 qt

2 pt

4 pt